# Paranormal Portal:
## Enter at Your Own Risk

![Tunnel with escalator and skull-like figure]

## Igor Kryan

ISBN 978-0-359-13417-5

www.youtube.com/c/IgorKryan

Printed in the United States of America

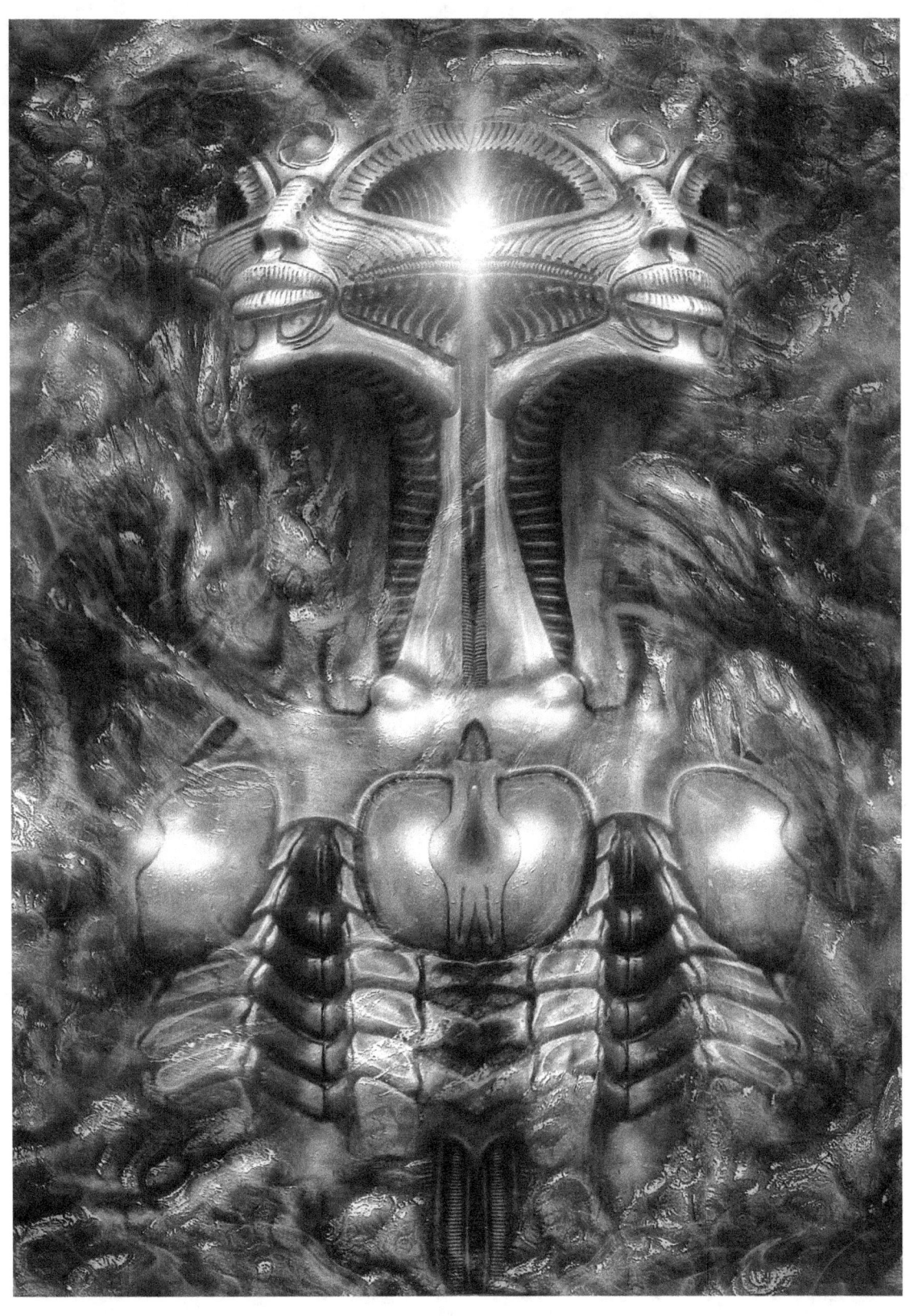

Ghost Encounters, Alien Abductions and Unexplained Happenings

## Intro

Are all these stories about ghosts encounters, alien abductions and paranormal happenings just urban legends and Halloween tales or are they real? Today you shall find out the truth. Yes, they all real. And today you shall hear the scariest and the weirdest of them all.

Authentic paranormal stories from very respectable people like doctors, teachers, officers and even queens and presidents. These encounters are guaranteed to be genuine because people telling them have nothing to gain and a lot to loose: public ridicule, professional licenses revoked, sanity questioned and firing from their important jobs. I was collecting these supernatural stories for over 33 years so read them at your own risk.

Content:

## Chapter 1. Imaginary Friends

My physician Lena lived in Saint Petersburg, Russia (former Leningrad). When she was 4 or 5 years old she started to see a little girl in her room. That little girl was telling her that her name is Veronica and she is very hungry. So Lena was running to the kitchen to bring her food. Every time Lena would return to her room Veronica would no longer be there. Lena grow up, finished medical school in UC Berkeley in California and thought it was just a child imaginary friend who was talking to her. About 10 years ago, she went to visit her childhood city of Saint Petersburg. There she met her former neighbor, now a very old lady well in her 80s. That old lady told Lena after Lena moved out, a rich family from Moscow of Baltic sea boat owners moved in to her flat, but before Lena was born and her family moved in, the city was under the siege by Germans, when another family lived there and that old lady was just a little girl and she was friends with another girl who lived in the same apartment. Old lady said that, unfortunately, that family starved to death during German siege and her good friend - that little girl did not make it. Girls name was Veronica. Lena returned to California with gray hair because she found out that her childhood friend was not imaginary at all. 10 years after she is still saying that encounter made her question everything she learned in UC Berkley and medical school about the finality of death.

This story comes from a teacher who took her sabbatical year and her 6 years old son and decided to escape noise of Silicone valley escaping to rural Montana for summer. The house was so rural it did not even had a bathroom inside but about 100 yards outside, in the field. One night the teacher steps outside her door. She looked though the window and saw her son going though the night field. Teacher asked son to wait for her, she put her dress and come out but there was no one in the restroom or in the field. She went back to the house and saw her son sleeping in the bed. She asked if he went outside. Son said, "No mommy, but some boy comes to me every night. He tells me the stories how he was snatched by a wolf from this farm." Mom said, "We need to report this lost boy to the police." And son said, "No need mom, it was not in out time but over 100 years ago." Mom and son were on the next flight back to California.

Another retired teacher told me a story once. She and kids were in summer camp settlement in Oregon. One night a girl opened the window and started climbing out saying that boys were calling for her to play in the forest. Teacher said it's night time and time to sleep not to play. Next year, teacher was with another class and caught another girl climbing out from the same window. When teacher asked that girl to explain herself, she said that two boys came to the bungalows and were calling to play in the forest at night. It was all girl summer camp and the nearest settlement was 30 miles away.

Boy's mother and father told me a story that they were picking up their 5 year old from a kindergarten and were 2 hours late. So no one except a security guard and the kid were there. They asked they son if we was not scared being almost alone in the kindergarten. He said, "Not at all, and I was playing with the girl." Parents asked if there are other parents were so late to pick up the girl and the child said, "No, that girls always stays there and she lives there." Next day parents came to thank the teacher for telling security guard to watch the child and wait for them. They also asked about the girl who their son claims to stay overnight in the kindergarten. Teacher became pale and said he is the third child who made such claims, and there was unfortunate accident when a girl ingested something several years ago and was not saved in that kindergarten.

San Francisco Nurse came from the night shift and saw her 4 years daughter is not sleeping but playing and laughing in the corner of her room. She asked why she is not asleep, the girl said, "I am playing with my grandpa." Nurse, knowing that grandpa passed away before the daughter was even born asked, "What grandpa?" Daughter said, "That one in the white hat" and point the finger into the empty space. Nurse was till shacked the morning knowing that her father loved his white cowboy hat.

Parents were camping in upstate New York wilderness on the way to Niagara falls. At night all their 3 kids disappeared from RV camper and in the morning they were found sleeping on the field nearby. Parents were questioning kids saying they were worried sick about them. All 3 kids said they were playing with a blond boy in the field at midnight. Locals, who were helping parents to find the missing children, said that field used to be an old graveyard.

UC Berkley student rented a room and send a snap-chat video of the room to his friend. Friend instantly replied, "Your new roommate is cute." Student asked, "What roommate?" Friend replied, "The girl on the sofa." Student was shaking and said, "I am alone in the room and have no roommates."

## Chapter 2. Queens and Presidents

Some of you might think that even though the witnesses of the paranormal stories seems to be officers, teachers, doctors and other respectable people but why would not the president of the United States come forward and say something. In fact he did. Nearly every President saw Lincoln ghost in the white house. President Truman stated that he clearly saw Lincoln in his bad with his wife. Than he changed the story saying that Lincoln was alone in his bedroom to make it sound more sane.

Nancy Reagan used to assemble entire ghosts summoning ceremonies in the oval cabinet claiming that it is the best place to summon ghosts because of its rounded architecture. After spiritual séances she would convey the advices ghosts gave to her to President Reagan, who after his assassination attempt in 1981 would never do anything not consulting with Nancy. For 8 years Reagan followed everything ghosts were telling him to do though Nancy.

French Queen Mary Antoinette Syndrome or sudden whitening of hair and sometimes eyes after seeing something profound. Hundreds of witnesses saw how Mary Antoinette brownish hair turned stark shining white in minutes white after following the ill-fated Flight to Varennes during the French Revolution. This paranormal fact was documented in all French history books and has no rational explanation. It's estimated that over 10 Million humans around the globe had the same experience. We all have a friend or co-worker who has a white strand of hair. And if you ask how it happened - he or she will likely answer - within seconds after the profound shock.
I had a friend who was a paranormal researcher. 6 years ago he went to Chernobyl 30 kilometers radioactive zone. When they found his body there, his own parents could not identify him - his dark hair and brown eyes turned stark white. It took fingerprinting, tattoo and DNA analysis to positively match him. How a young dark looking 25 years old man could turn into an elder in 1 day with white hair and eyes without anything paranormal involved?

A story from modern days was recorded by me initially told by a very famous political person who wishes to remain anonymous. He said he was visiting a castle in England. The castle was closed but he met a guide dressed in old English clothes who showed him the castle. Once they entered the dungeon his guide disappeared like he dissipated in the air. Being rich and important person he came to the administration and complained about the silly joke and circus trick of that guide, asking if they want to cause an international incident. The administration told him that on Monday castle is always locked and no one is allowed there and they not sure who it was since there are only two guides with the keys and they both women.

U.S. House representative Bettina Rodriguez Aguilera has a long list of accomplishments to bolster her campaign in Florida.

However, she is, perhaps, best known for claiming that she was abducted by space UFO aliens as a child. Rodriguez Aguilera says she was taken aboard a spaceship as a young girl by blond extraterrestrials who resembled the Christ the Redeemer statue in Rio de Janeiro. She said they told her that the "center of the world's energy is Africa" and that thousands of humanoid skulls will be discovered in a cave in Mediterranean and that will change human history. You can discard her statement as political campaign trick, but the problem with it, she was maintaining them since the age of 7.

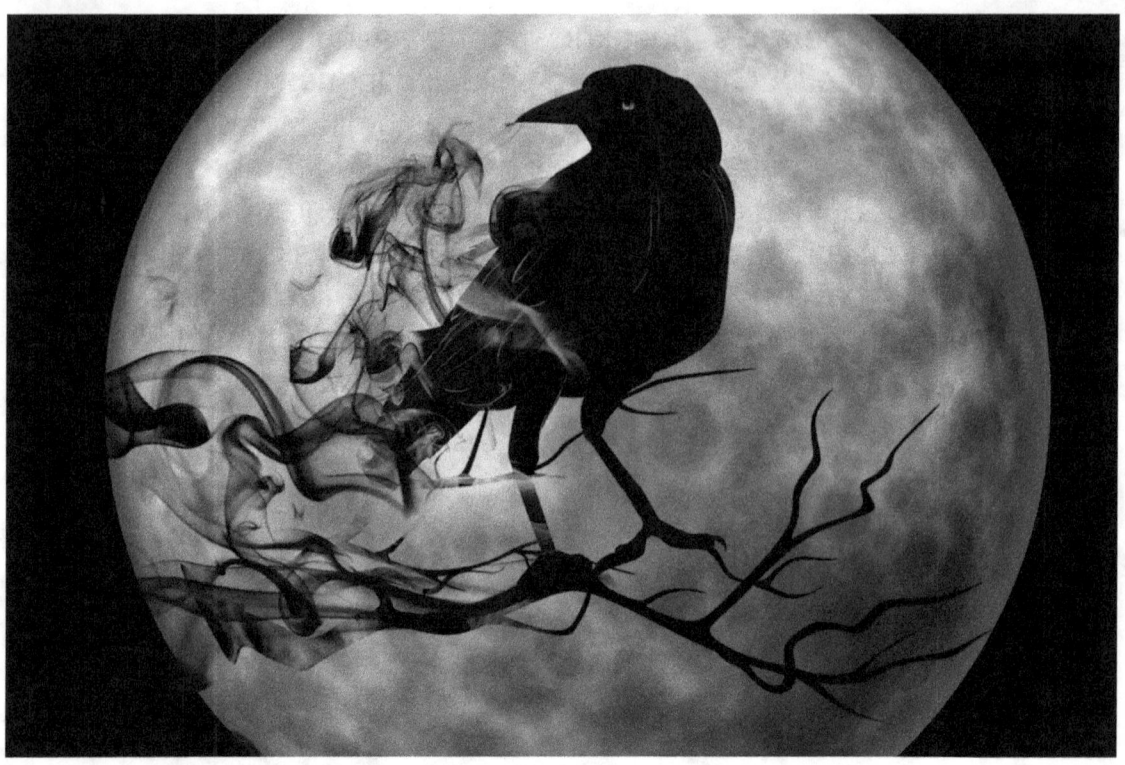

## Chapter 3. Raven and Death Valley

I remember when I was a child I liked to walk with my friend around the block and there was a creepy aging house with an old man living there with his pet raven. We always would stop by to take a look at the raven. One day we swear we heard the raven say, "Help, evil witch turned me into a bird" - we run away home in fear. Next month we drunk some beer to be brave - in Ukraine in 1990s beer was not considered an alcoholic drink and every child could legally buy it. And we came to that house again and the raven said again, "Help, evil witch turned me into a bird." We were just standing there paralyzed with shock with our eyes bigger than plates, our jaws on the ground and our hair raised. Then creepy old man showed up looked at us and said, "It took me 5 years to teach him say that to ward off wondering kids like you." Next year old house was demolished and old man and his raven moved out and I never saw or heard from them ever again, but I am sure there are many people in Europe with gray hair because of that raven.

About 15 years ago I and my friend decided to visit Death Valley in California. We stopped at another friend's place, who was a retired police officer in the city of Trona. Then we decided to go stargazing and, if we lucky, to spot some alien activity because locals said they saw strange green lights zipping the sky a night before. As night falls, we started driving via a dirt road and then all 3 of us saw a young black woman covering her head with some cloth standing in the middle of the road. Instead of stopping and picking her up, this former cop just accelerated and passed by her. I asked, "Maybe she is in trouble. We need to stop. You are an ex cop, we should help." Cop looked at me an said, "If we pick her up, I will be the one needing help. Last recorded radio transmission from my partner to dispatch 5 years ago was - I just pick up a badly sunburned woman from the side of the road. She needs medical attention. That was the last transmission he ever send. Next day they found his police cruiser abandoned and he was never seen or heard from ever again. Still a missing person." Cop added: "2 years later I found sheriff documents from early 1900s saying that young woman riding a horse ventured too far from the farm, fell from the horse in the desert and deceased from sun and heat exposure. That woman was not black - she was white, sun made her black 100 years ago."

## Chapter 4. Cats and Ghosts

The cold war was at the end but still raging and in the basement of our 10 story apartment building we had a nuclear fallout shelter in case the cold war would turn hot. Neighbors and my family were using compartments there for storage. I was just 11 years old or so when I was send there to bring some potatoes from the basement. I really enjoyed how my grandma fried them - no one was able make potatoes as delicious as she could. When I was in the basement the power outage hit the city. They become very common before the end of the Soviet Union in early 1990s. So I had my big metal Soviet flashlight with me. But you know, the Murphy's law - flashlight malfunctioned too. I was standing in the dark in the middle of that bunker when something jumped on my head. Next thing I felt - sets of sharp nails on my face and head. I decided to hit whatever jumped on my head with that large metal flashlight - only to hit my forehead and to fall on the floor. I woke up about 5 minutes later thinking that I am dead - it was just darkness and silence around me. I was thinking - its really sucks being dead. That monster must have killed me. Neither angels around, nor even hellish creatures of the underworld - just darkness and silence. Then I saw a pair of green eyes staring at me from the heart of darkness. I stood up on my feet and run as fast as I could. Then I realized I am not dead - just running inside that dark bunker. When I stopped my heart was racing 300 times a minute and I had tremendous headache and few scathes on my temples. Next day I was sick and the day after I gathered my friends and we went to the bunker armed with every possible medieval weapon we could find: from a torch to a spear. All we found there was a mommy cat and a few newborn kittens. I must have ventured too close to her kittens den and she jumped on me to protect them 2 nights before. We brought them some milk and my friends were laughing at me for a long time saying - so pussycat attacked you, next time a mouse will attack you - bring an army.

Now hear some real paranormal story about cats.
Another family I knew lived in New York. One night their 5 year old son with scathes on his forehead and cheeks was crying and trying to wake up his parents. It turned out that the gas pipe in the old stove cracked and gas was coming into the apartment. Parents shut down the gas, thanked their child but asked how did he got scratches on his forehead and cheeks. The child said he was sleeping but then something like a big cat jumped on his chest and started to scratch his face, so he woke up and "the air in the room smelled bad." There were no cats or any other pets in the apartment at that time.

## Chapter 5. Dogs and Ghosts

February 27, 2010. I was with my girlfriend in northern Chile Pacific beach - south of Peru. I went to explore beach caves while she was sunbathing. February is one of the warmest months there - since the seasons are inverted in Southern Hemisphere. In the cave, I discovered an old scary dog. I had some cookies in my bag and I gave her one or two. The dog ate cookies but instead of begging for more she run away from the cave and looked at me. I stepped out the cave and 9.0 earthquake hit Chilean coast. One of the strongest in the recorded history. The cave has collapsed. That dog somehow saved me.

Maybe it was a payback for the time when I saved the dog on March 27, 1989. I spotted a puppy on the iced freeway in Kiev, Ukraine. The moment I bent to lift the puppy up I got hit in the head by a passing truck and fell on the ice and snow. I don't remember any of it, my friend told me about it weeks later, when I woke up from coma in Kiev children hospital. They say you don't feel anything when you are in coma but I am sure that I saw how the spring was coming, first leafs appear and trees begin to blossom outside the window while I was in coma. When I woke up, I discovered that I lost much of my eyesight due to the severe hit in the vision brain cortex and skull fracture. My eyesight was surgically restored only in 2012. However, I also discovered that I have new talents that I never had before like drawing, painting, writing and, of course, story telling.

I witnessed the somewhat similar story that happened in my school in Ukraine, Kiev. Everyone was on alert because schoolboy did not return home from the school the previous day. People disappearance was very common in early 1990s in Ukraine. The next day after school the same pupil just ring his door bell. Worried sick parents and police questioned the boy but he thought that the day was yesterday and he returned from school to his home right away as usual. Even a year after, he still could not account for any of the missing time. His parents reported that boy has changed and became like new - his old scar disappeared and he started to receive better grades in school and no longer needed his glasses and got a dog.

The final story of the chapter. One Russian officer said that his commander decided to test his artillery system on the stray dog walking the military field. He aimed the cannon and pull the fire bar. Weapon misfired killing the commander. The same weapon never misfired before for 20 years. When military police engineers where checking the weapon after the incident, they could not find anything wrong with it and commission it back to the normal use - the artillery system was working for another 10 years. Since then that officer believes that there is a god who protects dogs and so should you.

## Chapter 6. Welcome to My Dreams

Construction worker told me that his sister always complained that she had a lot of nightmares, but she would never say what her dreams were about. One day while they were in the shopping mall he turned to talk to his sister, and almost had a heart attack because standing behind her was a man in old-fashioned clothes holding one of his sisters shoulders and looking at him with a very angry expression. She saw the shock on his face right away and shouted at him asking what was wrong. When he told her what he had seen, she started crying and said, "you just described the man that tries to kill me almost every night in my nightmares."

Boy's father woke up hearing the singing in the baby monitor to his child. He came to baby's room and hearing his wife singing though the baby monitor, "sleep baby, sleep baby, sleep forever." He came to wife's room to ask why she sings such a strange song and that moment his cell rings - it was his wife saying that she did not want to wake baby up and went to Walgreens 24 hours store. She was not in the house at all to sing that song.

In 2018 San Francisco Bay Area started to suffer the worst traffic ever. And two female friends rented an apartment near Google where they worked in Santa Clara to avoid the commute. On Friday night one of the females had drinks in the local bar and decided to go back on foot rather than wait for Uber to clear up her head. Once she was walking the alley, she felt that someone was watching her. She started to run and heard someone running after her. Finally, she got to her apartment and lock the doors and went straight to her bed. Next morning her roommate told her that she had the weirdest dream - she had dreamed that there was someone following first girl very closely, wanting to get into the house. And so she had placed the chair at the front door to prevent that person from entering.

Big rig driver said he signed to drive across the country after his beloved wife passed away to take his mind away from the loss. One day he needed to leave at 5am but was drinking in the bar the night before to ease his pain. At around 6am he fell asleep driving on the highway. He woke up a few minutes later when someone was petting his head and whispering into his ear ,"We will be together again but it's not your time yet." Then he saw his truck perfectly parked on the side of the highway with his keys in his pocket. He run off from his truck and was shaking for an hour trying to understand what just happened.

NASA engineer had a college friend, who just vanished one day and no one ever found him. One night his friend appeared in engineer's very vivid dream. In his dream engineer asked why he had disappeared. This friend told him: "Sorry man, love of my life, my girlfriend Sophie cheated on me. I just could not take it and jumped off the bridge." Once engineer woke up, his first thought, it was very vivid but just a dream. The next moment their common friend Sophie called his cell phone, who was not calling before for years.

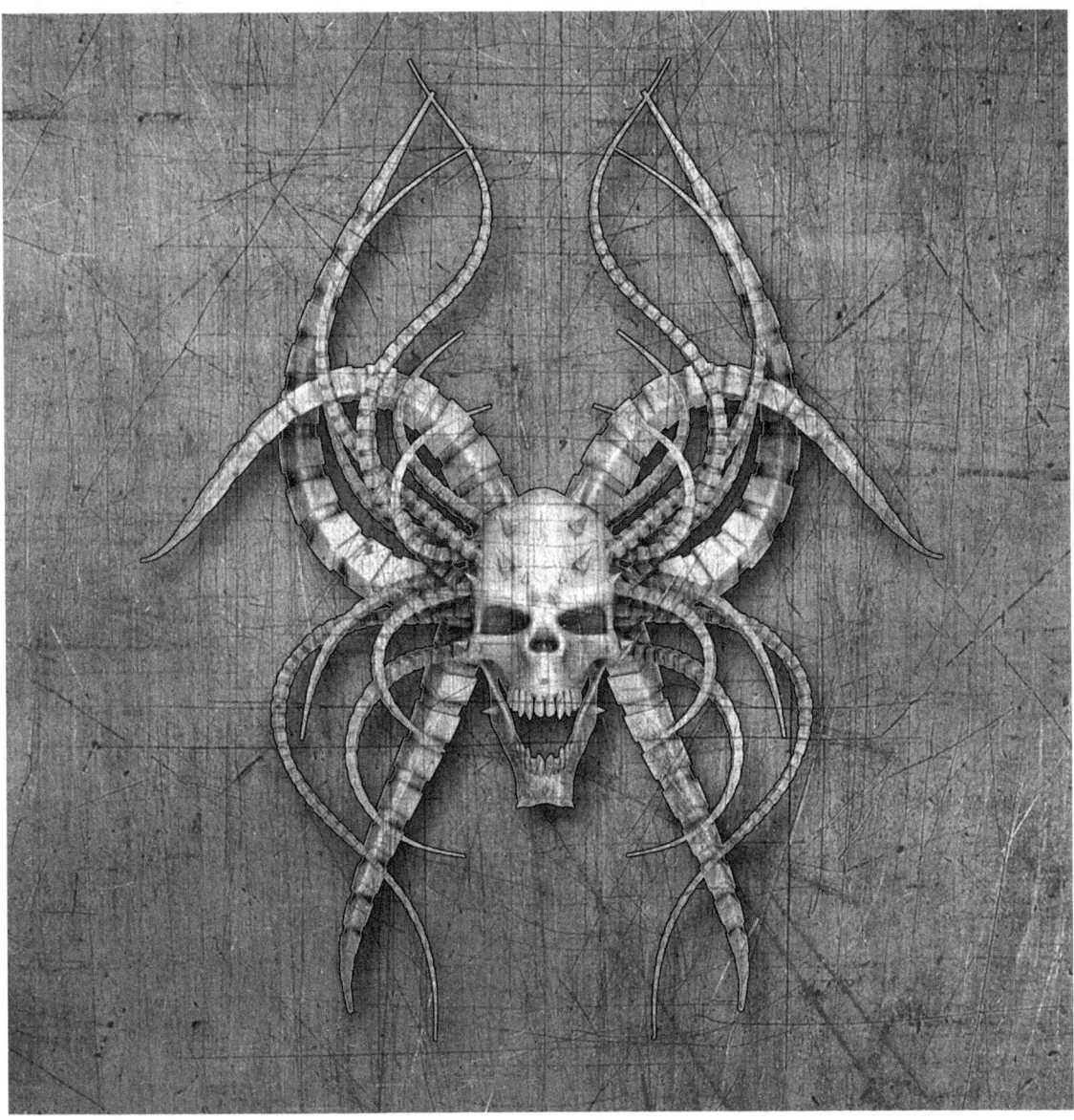

## Chapter 7. Stories from the Graves

This story was told by a former teacher, now a grandma. She had a dream. Her deceased husband came to her in the dreams and told her that his head begin to freeze. She went to the cemetery and find out that the tomb stone was moved for about a foot in the place where is her husband head was buried.

Do you know that only 5% of all people ever saw a UFO but over 50% or 4 billions of people saw a ghost. By the way, that's includes me I saw the host of my dog very clearly on the 3rd day she passed away. After the sighting I was happy and get a sense of relief.

But it's not the case of many others who saw ghosts. And don't tell me that 4 billion people including kids and very well educated folks who saw ghosts are either lying or drunk. The device was build by the hands of Nicola Tesla to detect paranormal slightest electromagnetic vibrations known. It is analog but far more superior and complex to anything that was ever built after him. We tested it in the only known California necropolis called Colma and it was showing significant reading there but went mostly silent in urban environment even around microwaves and cell phones. The reading were off the chart in one particular place of the graveyard. Old cemetery security guard told us that many years ago a young girl who lost her parents was buried there by her uncle but since then every night she is looking for them. He said that torn electrical wires and broken plants are very common in that area of the cemetery.

## Chapter 8. Churches, Priests and Prayers

This story was told by the United States veteran. After September 11, he was summoned to go to the war in Afghanistan. On the way to his regiment he spotted a small church with blue roof near the side of the road. And he stopped there to pray for his family, wife and kids and for his own safe return home. In 1 year he returned and decided to come to the same church to thank God for his safe return to his family. He could not locate that church anymore, so he came to the local eatery and start asking about the church but no one knew about it. Finally one old lady said him, "Son, the church with blue roof you are describing indeed was here and I attended Sunday prayers there it when I was a little girl but it was burned down 50 years ago."

The priest in another church told me this story. He was praying late in the church when two children in the hoods knocked the doors. He opened the doors and asked the children if they are lost. Children did not answer but entered the church and removed the hoods. They eyes were pitch black. They asked the priest for water. He gave them holy water. They took it but did not drink it and left.

A female linguist I knew once told me an incredible story and even showed the video. She said she never told anyone about it but here it is: When she was a very little and could not even talk - her father videotaped her standing looking somewhere and doing baby talk. She grow up, learned 5 languages, got Master's degree and had her own child. She decided to show her child how she was as a little baby. However, those tapes her dad made required a VCR that she did not have so a friend transferred them to digital format for her. She turned the tape on and was shocked to the core - her baby talk was not a baby talk at all - she was speaking pretty clear Japanese where she was praying and begging Japanese god Inari Okami to save her husband in the battle because she is pregnant with his child.

## Chapter 9. Unborn

One female attorney lost her personal iPhone. She had work phone in her office and her personal cell phone on the speed dial. She dialed her phone twice and both times some little boy answered: "Mommy you should not have had an abortion." She thought its some sort of sick jokes of her coworkers, since she really had an abortion a couple of years ago. In the end of the day she returned to the car and found her phone with 2 missed calls from her office phone. The car was closed and fully alarmed.

Another mom told me that her 4 years old son told her that he knows when he was in her stomach mom was sick and in the hospital for a long time and crying a lot. Mom was surprised since she never told him about that. So mom asked if her son knows where kids come from when they born. Her son told her, "Yes, mommy, they are brought by a big bird as you told me but I am special, I was not brought by a bird, I was in your stomach."

This is the story a teacher told once to the class. When she was 8, she used to wave at a man who passed by her window every morning at the same time. A couple of times she tried to run stairs and meet a man but was always somehow missing him. Once she decided to wait for him on the street. He appeared and shook her hand. Next moment her mother caller for her asking her why she was outside and she told her that she was meeting the man. When she starting describing that man - mother started crying and said that the person she was describing was exactly like her first love who passed away - the father of the daughter he never saw.

Mother was putting his 4 years old son to sleep and asked if she should turn off all lights or leave a night light. Toddler said to his mom, "It's no difference, the darkness shall not disappear." Mother asked, "What darkness?" Child answered, "It was dark when I was inside you."

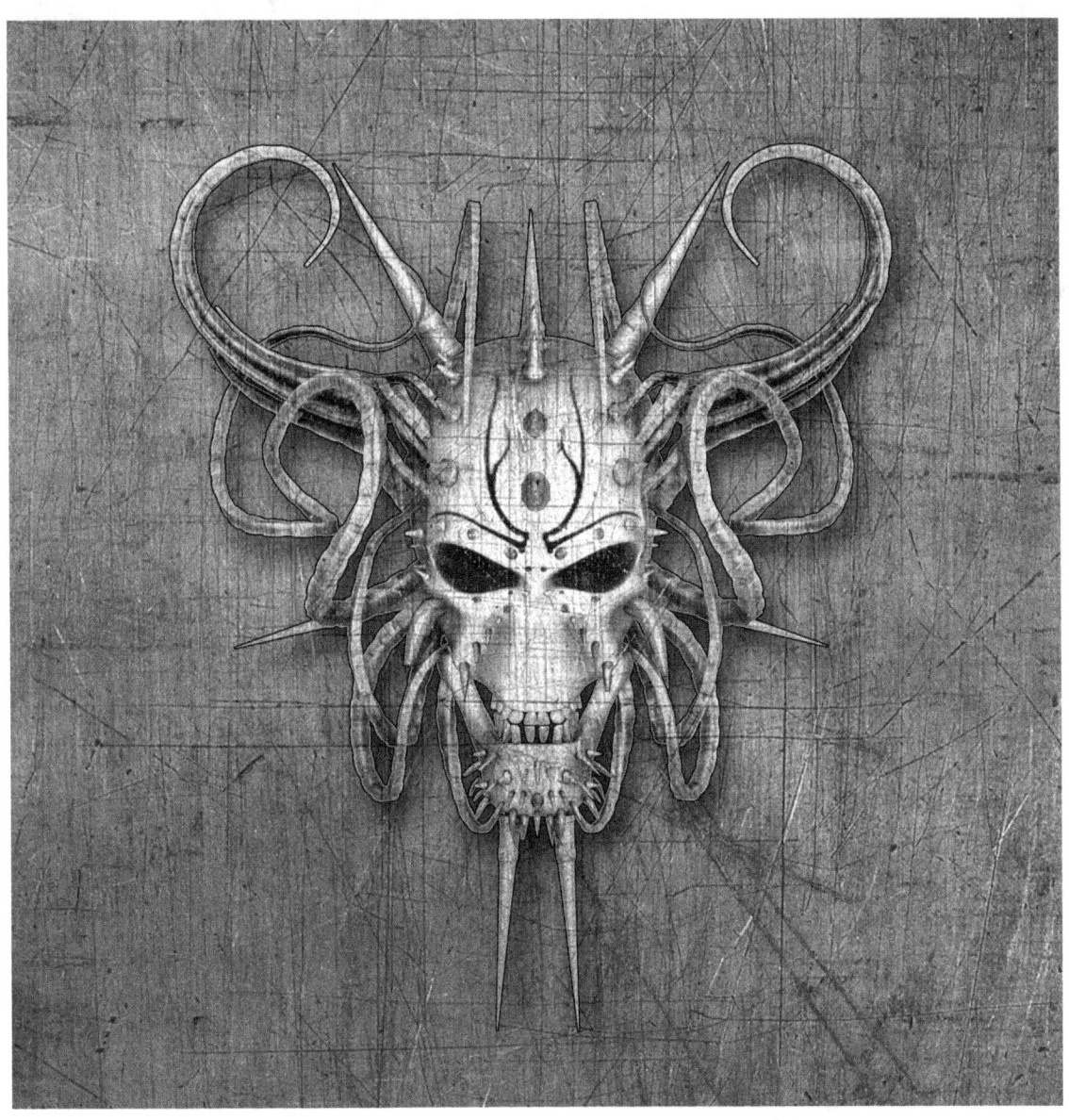

## Chapter 10. Nonbelievers

Some of you might say that we all adults and don't believe in any supernatural stuff. Aren't we? Everything was debunked like 1980s San Francisco Hospital story when every Friday at midnight a patient died from room number 13. So two doctors were left overnight in room number 13 which was an ICU on Friday with a patient to see what will happen. At exactly midnight janitor came and plug the vacuum unplugging patient breathing apparatus and started cleaning.

My friend went to the Gypsy fortune teller and asked her about his future. She pull out tarot cards and look at his hands and said. I foresee a bright and fiery future for you. This card shows you in bright glowing helmet like a crown, and this that you quickly climbing the ladder of carrier or life, and this one showing you hold a grateful beautiful lady in your hands. He said, "No wonder - I am a firefighter."

But now let's hear some real paranormal stories from very respectable people like doctors, teachers, officers.

Ukrainian Medical University dean was sitting on the table hearing unexplainable stories that happened in anatomy section of the university. After hearing them all he said to his colleagues: "Nothing you told me is paranormal and supernatural. In fact, I lived for 60 years and never in my life I witnessed anything extraordinary that cannot be easily explained." At that moment all windows in is cabinet cracked or shattered. There was no high winds or any kind explosion anywhere in the university that day.

One nurse was walking her 6 years old son in the yard when she noticed a stranger staring at her boy from the bushes. She told the stranger, "Go way or I will call the police." Stranger said, "I will go but you need to protect your boy, don't take your eyes from him today." Disappointed, nurse went home. In 30 minutes her boy fell from 5th store balcony - he landed in the sandbox and was not seriously hurt.

San Francisco police force veteran was just a student when he lived on the very top floor of Park Merced high rise building near State University where he studied Criminal Justice. One day it was raining pretty hard in San Francisco and thunderbolt strike the tower apparently causing the elevator he was riding in to stuck. He started to push all buttons in the elevator and elevator begin to move again. Then it stuck again but opened the door between the floors. Future officer looked up and saw a giant humanoid ape face started sticking into the elevator cabin from the upper floor though the open doors. It was the only time he pissed his pants. He jumped to the lower floor though the open doors and then for 3 days refused to leave his apartment.

Another retired female police officer told a story to a common friend. Once she was late for her shift and was running to the car to get to the police cruiser in the precinct in Oakland, California. When she was opening car door - another creepy officer she never knew passed by her and said, "Turn off the soup on your stove." She stood shocked and remembered that she forgot to turn off her stove gas with the soup on top of it. She ran back home and turned off the stove. She was late for her work. When robbery in progress call was made she had to take another car but not her regular cruiser because her partner already left. Once she arrived on the scene suspects were apprehended but one of them managed put 3 bullets though her regular police car windshield into her seat and head reclining attachment. No one was hurt, since she was not there and was late. She still have no idea who that creepy officer was who stopped her.

## Chapter 11. September 11

Last year, when medical professionals daughter was just over 3 years old, they were watching a TV show about 9/11. It was on or around the anniversary of the event. Their daughter, who was drawing nearby, looked up when the screen showed a plane hitting one of the World Trade Center towers. She said to her parents: "I died there." Then she just went back to drawing like she hadn't said a word. They had never talked to her about the concept of death, and had never discussed 9/11 with her.

When a firefighter was 20 years, many years ago, he kept having dreams about a woman with long black hair named Aurora. They all were different dreams but for some reason, her distinct face and name always were present in them. It got to the point where he would wake up frustrated and confused, trying to Google search her name or find out how he was connected to her. After a few months later she stopped showing up in his dreams.

On 9/11 that firefighter was driving his fire truck going to twin towers fire with sirens. He about to pull out and merge onto a highway when the girl stood in front of his fire truck. She look exactly like Aurora from his dreams. Firefighter slammed the breaks and stepped out of his truck. The girl run away. Once they were approaching twin towers, the tower collapsed right in front of the truck. Firefighter picked up his phone to call his family to tell that he was saved and noticed several missed calls from an unknown number. He dialed back and the voicemail said: "Hi! This is Aurora, please leave your name and number." Later that day in chaos firefighter lost his phone but he is still thankful to that lady who somehow saved his life.

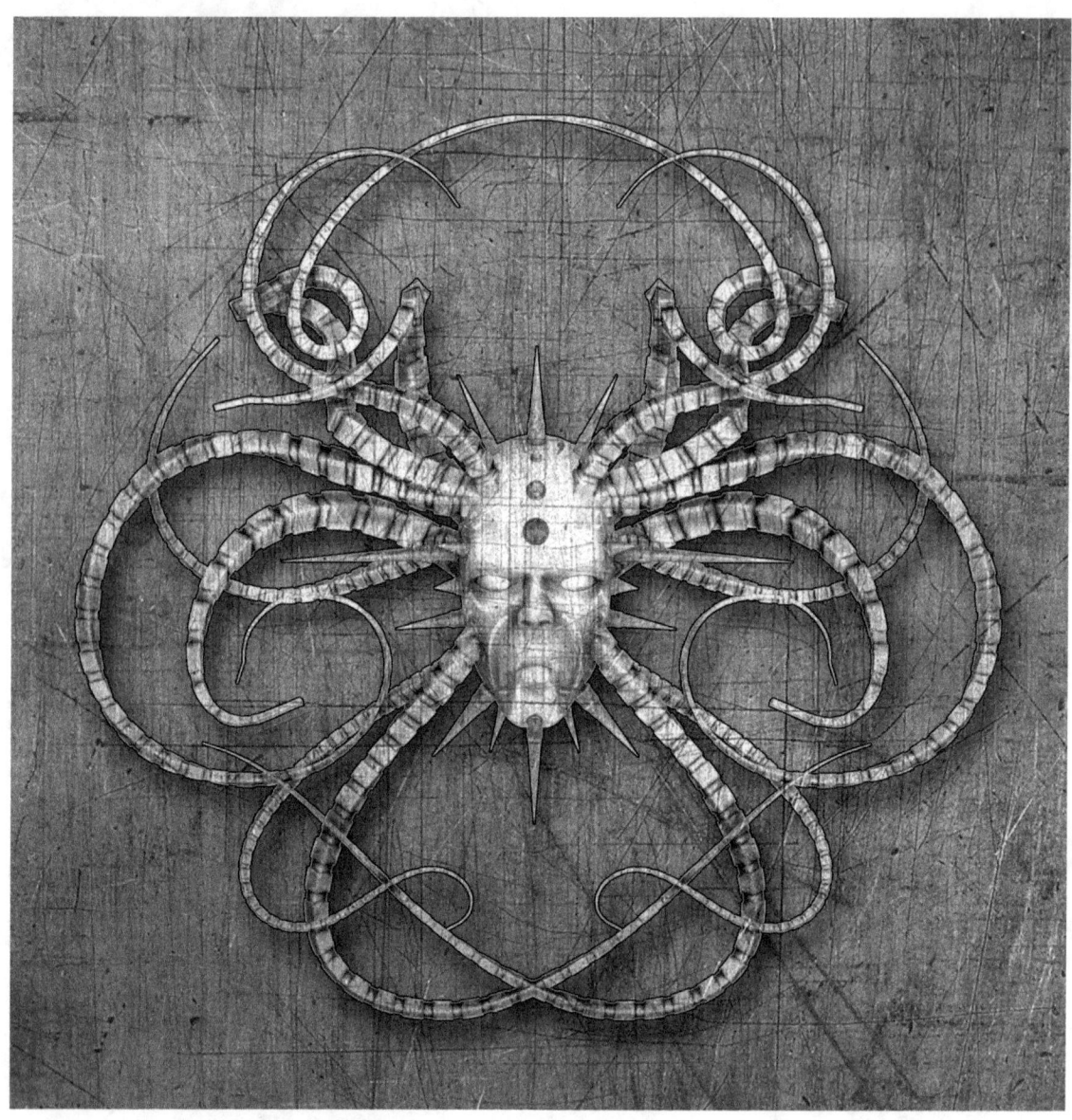

## Chapter 12. Paranormal Lightings

Believe it or not, but until 2014 most scientists denied that Ball Lighting phenomenon actually exist, despite numerous accounts from around the world. The first optical spectrum of a ball-lightning event, published in January 2014, included a video at high frame-rate.

One early account was reported during the Thunderstorm near church in Devon, in England, on 21 October 1638 following happened:  "Four people died and approximately 60 were injured when, during a severe storm, an 9-foot (3 m) ball of fire was described as striking and entering the church, nearly destroying it.

Large stones from the church walls were hurled into the ground and through large wooden beams. The ball of fire allegedly smashed the pews and many windows, and filled the church with a foul sulphurous odor and dark, thick smoke. The ball of fire reportedly divided into two segments, one exiting through a window by smashing it open, the other going though the walls of the church. The explanation at the time, because of the fire and sulphur smell, was that the ball of fire was "the devil" or the "flames of hell". Later, some blamed the entire incident on two people who had been playing cards in the pew during the sermon, thereby incurring God's wrath."

November 4, 1749 Admiral Chambers on board the Montague ship wrote: "I was taking an observation just before noon...I observed a large ball of blue fire about three miles distant from them. They immediately lowered their topsails, but it came up so fast upon them, that, before they could raise the main tack, they observed the ball rise almost perpendicularly, and not above forty or fifty yards from the main chains when it went off with an explosion, as great as if a hundred cannons had been discharged at the same time, leaving behind it a strong sulphurous smell. By this explosion the main top-mast was shattered into pieces and the main mast went down to the keel. Five men were knocked down. Just before the explosion, the ball seemed to be the size of a large mill-stone."

An English journal reported that during an 1809 storm, three "balls of fire" appeared and "attacked" the British ship HMS Warren Hastings. The crew watched one ball descend, killing a man on deck and setting the main mast on fire. A crewman went out to retrieve the fallen body and was struck by a second ball, which knocked him back and left him with mild burns. A third man was killed by contact with the third ball. Crew members reported a persistent, sickening sulphur smell afterward."

On April 30, 1877, a ball of lightning entered the Golden Temple in India, and exited through a side door. Several people observed the ball, and the incident is inscribed on the front wall of Darshani Deodhi.

On November 22, 1894, an unusually prolonged instance of natural ball lightning occurred in Golden, Colorado, which suggests it could be artificially induced from the atmosphere. The Golden Globe newspaper reported, "A beautiful yet strange phenomenon was seen in this city on last Monday night. The wind was high and the air seemed to be full of electricity. In front of, above and around the new Hall of Engineering of the School of Miners, balls of fire played tag for half an hour, to the wonder and amazement of all who saw the display.

On May 22, 1901 in the Kazakh city of Uralsk in the Russian Empire (now Oral, Kazakhstan), "a dazzlingly brilliant ball of fire" descended gradually from the sky during a thunderstorm, then entered into a house where 21 people had taken refuge, "wreaked havoc with the apartment, broke through the wall into a stove in the adjoining room, smashed the stove-pipe, and carried it off with such violence that it was dashed against the opposite wall, and went out through the broken window."

On April 29, 1925 in Germany multiple witnesses saw a silent ball move along a telephone wire to a school, knock back a teacher using a telephone, and bore perfectly round coin-sized holes through a glass pane. Over 700 feet (250 meters) of wire was melted, several telephone poles were damaged, an underground cable was broken, and several workmen were thrown to the ground but unhurt.

Hundreds of well trained military Pilots in World War II described an unusual phenomenon for which ball lightning has been suggested as an explanation. The pilots saw small balls of light moving in strange trajectories, which came to be referred to as foo fighters. According to almost all pilots who saw them - the balls seemed to be following the planes and exhibiting some sort of either intelligence or at least formations and strategy.

Submariners and NAVY officers in the Second World War gave the most frequent and consistent accounts of small ball lightning in the confined submarine atmosphere. There are multiple accounts of repeated inadvertent production of floating explosive balls. Official NAVY explanation was: "When the battery banks were switched in or out, especially if mis-switched or when the highly inductive electrical motors were misconnected or disconnected - the balls are produced."

However, attempts later to duplicate those balls with a surplus submarine battery resulted in several failures and an explosion.

And this is the more recent statements from Airline Pilots and passengers to FAA "We seated near the front of the passenger cabin of an all-metal airliner (Eastern Airlines Flight EA 539) on a late night flight from New York to Washington. The aircraft encountered an electrical storm during which it was enveloped in a sudden bright and loud electrical discharge (00:05 h EST, March 19, 1963). Some seconds after this a glowing sphere a little more than 20 cm in diameter emerged from the pilot's cabin and passed down the aisle of the aircraft approximately 50 cm from me, maintaining the same height and course for the whole distance over which it could be observed."

Finally, December 15, 2014, flight BE-6780 (Saab 2000) in the UK experienced ball lightning in the forward cabin exiting the aircraft and producing lightning striking the aircraft nose.

Stable considerable size and long lasting Lighting balls were produced in laboratory only once by Nicola Tesla. He could artificially produce over 1.5-inch (40 mm) balls and conducted some demonstrations of his ability, but he was truly interested in higher voltages and powers, and remote transmission of power, so the balls he made were just a curiosity for him.

When we were children we decided to reproduce Nicola Tesla experiments and capture "a living thunder ball." During huge thunderstorm we constructed 30 feet (10 meter) long lighting rod attached to the large 3 feet (1 meter) sphere made with mesh wire and placed on top of the building under construction on the high hill in the city of Kiev. Knowing that ball lightning's attracted to electric fields, we also connected 380 Volts 15 Amp power cable from the transformer to our tower and waited nearby in the same building for lighting to strike. When it finally happened, the lighting destroyed the entire tower along with the transformer making the wires glow for a few seconds but no lighting ball was produced to our disappointment. Later on I realized, that had we succeeded to produce 1 meter or 3 feet thunder ball, I wound not probably be around to write this book anymore.

## Chapter 13. Astronauts and Aliens

Many astronauts reported seeing those life forms in the space similar to foo fighters reported by Allied and German Air-force during World War II. One of them was original Mercury Astronaut Major Gordon Cooper and the last American to fly in space alone. On May 15, 1963 he was launched into space in a Mercury capsule for a 22 orbit journey around the world. During the final orbit, Major Gordon Cooper told the tracking station at Muchea (near Perth Australia) that he could see a glowing, greenish orb ahead of him quickly approaching his capsule. The UFO was real because it was picked up by Muchea's tracking radar. Cooper's sighting was reported by the

National Broadcast Company, which was covering the flight step by step, but when Cooper landed, reporters were told that they would not be allowed to question him about UFO sighting.

Ten years earlier, in 1951 Major Cooper had sighted a UFO while piloting an F-86 Sabrejet over Western Germany. "They were saucer-shaped glowing discs at considerable altitude and could out-maneuver all American fighter planes. Major Cooper also testified before the United Nations: "I believe that these extra-terrestrials are visiting this planet from other planets... Most astronauts were reluctant to discuss UFOs. I did have occasion in 1951 to have two days of observation of many flights of them, of different sizes, flying in fighter formation, generally from east to west over Europe. For many years I have lived with a secret, in a secrecy imposed on all specialists in astronautics. I can now reveal that every day, in the USA, our radar instruments capture objects of form and composition unknown to us. And there are thousands of witness reports and a quantity of documents to prove this, but nobody wants to make them public. Why? Because authority is afraid that people may think of God knows what kind of horrible invaders. So the password still is: We have to avoid panic by all means."

"I was furthermore a witness to an extraordinary phenomenon, here on this planet Earth. It happened a few years ago in Florida. There I saw with my own eyes a defined area of ground being consumed by flames, with four indentions left by a glowing flying object which had descended in the middle of a field. Beings had left the craft (there were other traces to prove this). They seemed to have studied topography, they had collected soil samples and, eventually, they returned to where they had come from, disappearing at enormous speed... I happen to know that authority did just about everything to keep this incident from the press and TV, in fear of a panicky reaction from the public."

NASA astronaut Story Musgrave claims to have seen glowing eel-like tubes swim through space. In the interview above, he explains that he saw this "creature" on two separate occasions. While some immediately dismiss this as space junk, possibly some type of hose that detached from a spaceship, Musgrave is insisting that the white eel had its own propulsion technique.

Major General Vladimir Kovalyonok was part of a crew manning Salyut VI space station in 1981: "When I was working at the Salyut orbital station, I saw something strange in a porthole one day. The object was the size of a finger. I was surprised to see it was an orbiting object. It was hard to determine the size and the speed of an object in space. That is why I can not say exactly, which size it actually was. [My partner Viktor] Savinykh prepared to take a picture of it, but the UFO suddenly exploded. Only clouds of smoke were left. The object split into two interconnected pieces. It was reminiscent of a dumb-bell. I reported about it to the Mission Control immediately."

In 2005, Leroy Chiao was commander of the International Space Station. While on a spacewalk, Chiao saw white lights aligned in an upside-down check formation whiz right past him. Some people have posited that a string of fishing boats along the South American coast could explain what he saw, but Chiao was 230 miles above Earth when this happened. Those would have to be some impossibly strong boat spotlights to be seen from all the way up there.

In 2014, European Space Agency astronaut Samantha Cristoforetti was on her way to the International Space Station for the first time when she saw the normally gray ISS was bathed with glowing orange light. Cristoforetti was taken aback by the beauty of what she was seeing, and in a blog post, she wrote, "The enormous solar panels were inundated with a blaze of orange light, vivid, warm, almost alien." But none of the other astronauts saw this effect before.

John Glenn who flew on the Friendship 7 spacecraft in February of 1962, suddenly noticed something strange outside his window while in orbit. He immediately reported to NASA that he was watching what looked like a group of little glowing fireflies dancing outside his window. During the early Apollo missions, astronauts reported seeing "light flashes". The crews of later Apollo mission were warned about this and reported that they also saw strange bursts of light, flashes and glowing orbs.

Astronaut Alan Bean allegedly saw glass domes from a long-extinct alien civilization on the moon. In an interview, Bean described the space as looking like "black, patent-leather shoes" from the surface of the moon. Hoagland maintains, "Space should be velvet-black. It should be inky-black. It should be infinity, unending, deep, endless black. It shouldn't be shiny as I saw it."

**Final Chapter: Really Weird Short Stories**

Spontaneous human combustion syndrome is so widespread it has even medical term SHC. When human burns from the inside while his or her clothes and surroundings remain in tact. Hundreds upon hundreds such cases were documented by police, medical doctors, coroners and government officials around the globe. In early 1990s when I was a teenager in Kiev one old lady was harassed by a local racketeer who was demanding an elderly lady on the market to pay a share from the profit to him. When she refused, he broke her merchandise. She said him that he will burn in hell tonight. At least a dozen of his gang members and other sales persons heard it.

Nothing happened that night, but a week after they found a pile of ashes in just slightly burned bed and clothes - all that is left of him. Any explanation to thousands of documented cases of spontaneous human combustion? In the church, I was told by a priest that is the way to send the demons back to hell.

Winter evening in Poland about 7:30 PM. Two young friends were building a snowman in the open field. While they were making snowballs, they both saw the field start to glow, like there is a bright light from inside. It started to dim, then builds slowly, then suddenly gets blindingly white and disappears. Then one of the friends look at the watch and realized it's 10:30PM. The other friend pager clock was also saying 10:30. So 3 hours seemed to pass in a few seconds for them, and the next day there were multiple large circles melted into the otherwise frozen snow field raging from 5 to 25 meters in the diameter.

Famous surgeon had unfortunate case of patient not making it though the surgery. He came home tired and fell asleep. He woke up at 12:57am because he had very vivid nightmare of that patient choking him in his bed saying how could he make his wife a widow. So surgeon went to the kitchen and drank some water. He came back to the bedroom and look at the clock which was showing 12:29am. He decided not to risk it and went to a nearby pub.

Highway patrol officer was driving down the road, about to make a right turn at a red-light. No one was next to him. As he was about to turn he heard crushing metal sound and terrible noises. He looked over and it looked as if the red car drove into a wall. There was nothing even near it. Car front end was completely smashed in. Like there was an invisible wall. Officer got out of the car to help the person. The red car was lightly smoking, or it could have just been steam, but no one was in the car. It was empty. That car was reported stolen or missing 10 years ago.

## About the Author

Igor Kryan was born in 1979 in Kiev, Ukraine. He arrived to the Unites States in 1999 and his daughter Alisa was born in 2008. The author graduated with Honors at the National Ukrainian Medical University (AA Degree in Health Science - 1999), San Francisco State University (BS degree in Biology - 2003), New York Vernell University (Master of Fine Arts - 2003), Amsterdam Ravenhurst University, Netherlands, (MS degree in Cell and Molecular Science - 2004), Canterbury University, UK and the Ukrainian National University (PhD Doctorate degree in Astrobiology - 2010).

Igor Kryan is famous in the United States and Europe for over 1,000 original artworks and about a dozen of published English books such as "The Source" in 2004, "Future History" in 2006, "History of the Impossible" in 2008, "Earth before the Pyramids" in 2009, "2012: Hoax or Shock?" in 2010, "What if the British had Won" in 2011, "Angels and Demons Art Trilogy" in 2012, "Creator's Riddle: Darwin vs. God" in 2012, "Apocalypse of Magdalene and Judas" in 2016 and "The Second Coming: Jesus Arrived but Government Hid Him" and "Alien Life and Dark Plasma: What Makes You Alive and Self Aware?" and "Paranormal Portal: Enter at Your Own Risk" in 2018 as well as several satirical books in Russian. Most of Dr. Kryan's books enjoy overwhelming success - hundreds of thousands electronic and digital copies along with thousands of actual paper copies were sold and many more are being currently sold across the globe.

Doctor Kryan is known for: Human and Animal Liberation Activism. Numerous books and publications. Professional paintings and drawings. Research of solar activity and UV radiation effects. Research of ancient and modern history and anthropology in conjunction with future analysis including prediction of 2008 Great Recession and New Cold War as early as 2005, Origin of life theory (2003), 50 consecutive miracles theory (2009), Universal Simulation Creator Theory (2012).

However, Doctor Kryan was not known for his association with CIA, FSB and ALF until he decided to go public in 2014 in order to save America from CIA sponsored modern day sophisticated enslavement and establishment of totalitarian state. The newest book CIA Trilogy: CIA Millennium Hilton (2013), CIA Earth Blood (2015), and CIA Oblivion (est. 2020) reveals both unknown author biography serving CIA, FSB and ALF interests and devious CIA plan to replace free world we love and cherish with totalitarian super state making us all obedient slaves in the process.

IK 2015